GLENVIEW PUBLIC LIBRARY
3 1170 00718 1757

D1606177

Carpenters

by Joan Freese

Lerner Publications Company • Minneapolis

Lerner Publications Company
A division of Lerner Publishing Group
241 First Avenue North
Minneapolis, MN 55401 U.S.A.

Website address: www.lernerbooks.com

Words in **bold type** are explained on a glossary on page 31.

Library of Congress Cataloging-in-Publication Data

Freese, Joan.
Carpenters / by Joan Freese.
p. cm. – (Pull ahead books)
Includes index.
ISBN-13: 978–0–8225–2800–5 (lib. bdg. : alk. paper)
ISBN-10: 0–8225–2800–2 (lib. bdg. : alk. paper)
1. Carpentry–Juvenile literature. 2. Carpenters–Juvenile literature. I. Title. I I. Series.
TH5607.F745 2006
694–dc22 2005005653

Manufactured in the United States of America
1 2 3 4 5 6 – JR – 11 10 09 08 07 06

Would you like to build a tree house?

Who can help you?

A carpenter can.

Carpenters build things in your **community**. Your community is made up of people in your neighborhood, town, or city.

Carpenters do important work. They build things out of wood.

Carpenters build small things like benches and birdhouses.

They build large things like houses and sheds.

Carpenters create new buildings and fix old ones.

Carpenters start work early in the morning.

Many carpenters carry **supplies**.

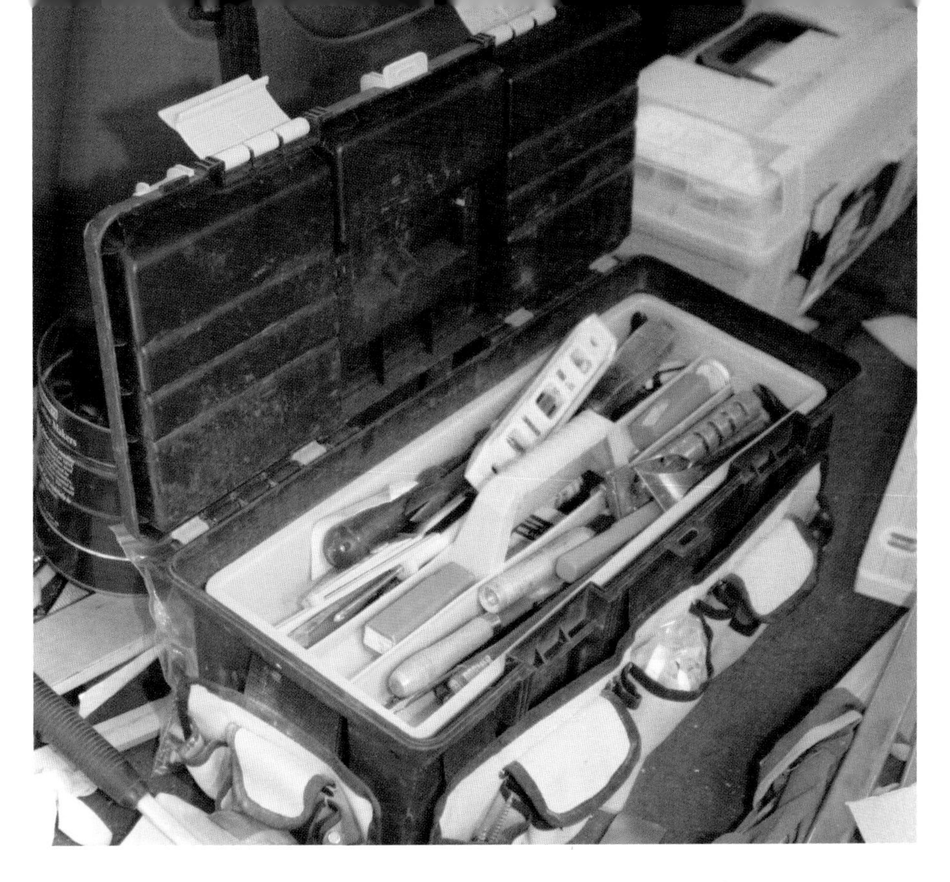

Carpenters need a lot of supplies. **Tools** are an important part of their supplies. Carpenters keep many tools in their toolboxes.

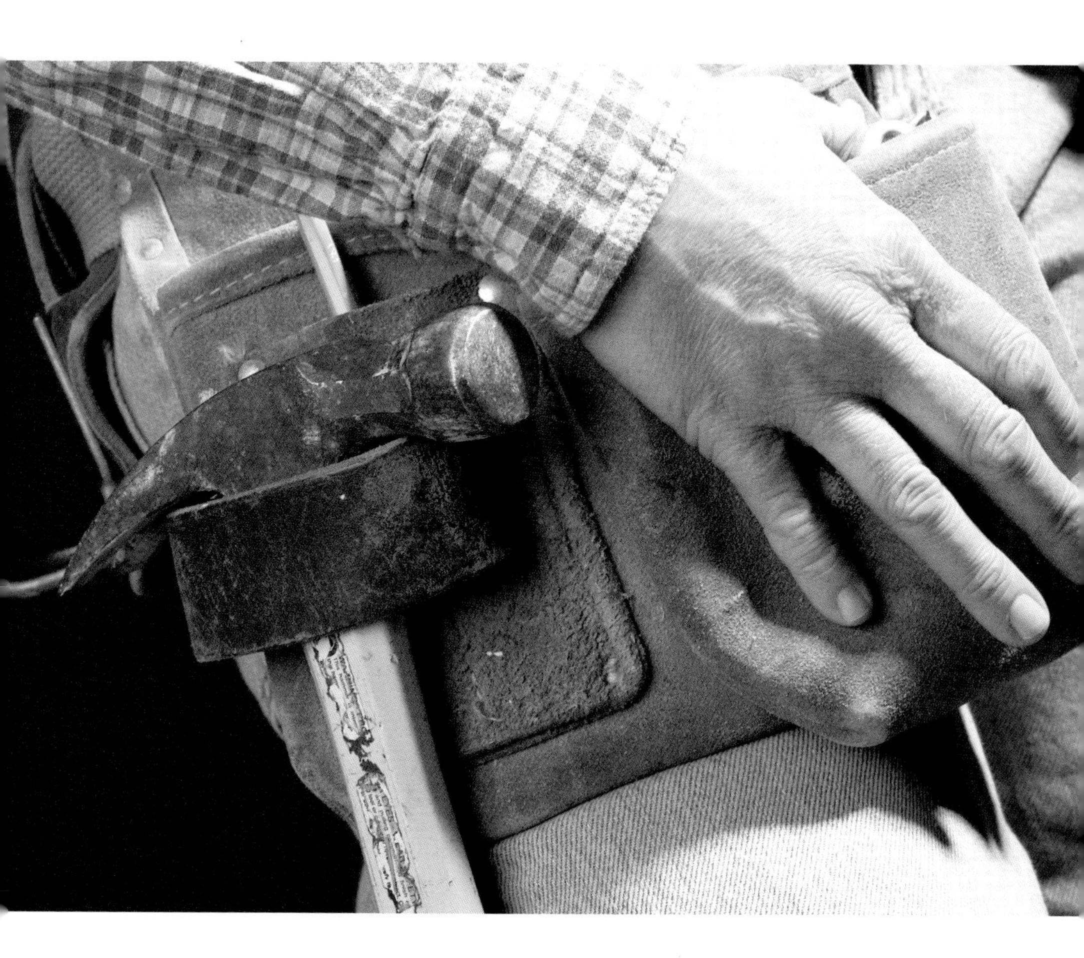

Carpenters carry small tools in their tool belts. What tools do carpenters use?

Carpenters
use saws
to cut wood.

Carpenters use drills to make holes in wood. Screws go in these holes.

Carpenters measure with measuring tapes. Measuring tapes help carpenters put wood together perfectly.

Carpenters say "Measure twice, cut once." This saying means "Be careful and plan ahead!"

Carpenters use **blueprints** to plan ahead. Blueprints show how buildings should be built.

Carpenters decide what supplies they need. Carpenters buy wood and supplies at a **lumberyard**.

Carpenters wear earplugs to keep their ears safe from loud noises.

Carpenters wear safety glasses to protect their eyes. Carpenters must be careful. What else must carpenters be?

Carpenters must be good at math.
They must measure carefully.

Some carpenters build furniture. They fix items made of wood.

Other carpenters help build houses. **Framers** build the frames of houses. Then finish carpenters add doors and woodwork.

Carpenters work with many people on the job. They work with **architects**. Architects design buildings.

They work with **plumbers**. Plumbers put water pipes in buildings.

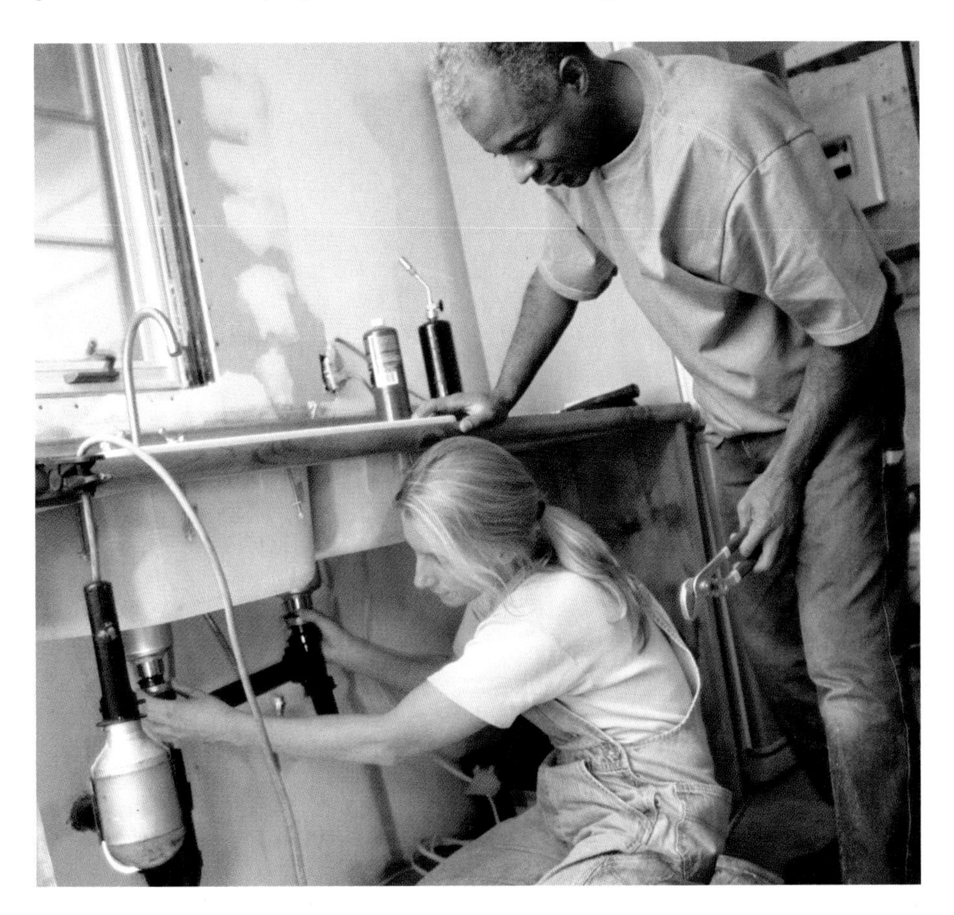

All these workers are very good at what they do. Together they make great things.

What would you build if you were a carpenter?

Facts about Carpenters

- Carpenters work on many different projects. They help to create buildings, bridges, and highways.
- Carpenters learn their skills on the job or in places called trade schools. Many communities have trade schools. Students there learn how to use tools and build safely.
- Carpenters must know local rules called building codes. These laws help keep buildings safe for people.
- Carpenters clean up the job site at the end of each day. The next morning they can start working right away.
- Carpenters must be good at math and science. They also must be creative and have good problem-solving skills.

Carpenters through History

- Carpenters used only hand tools until electric tools were invented. Electric tools made it easier for workers to build.
- People around the world have made things out of wood for ages. Spears, boats, and simple houses were some of the first things built of wood.
- Traditional Japanese houses are made of wood. Carpenters in Japan built very fine homes.
- The Vikings were skilled carpenters. These ancient warriors made big wooden ships that could sail great distances.

More about Carpenters

Check out these books and websites to find out more about carpenters. Or see if you can talk with a carpenter.

Books

Barton, Byron. *Building a House.* New York: HarperTrophy, 1990.

Burby, Liza N. *A Day in the Life of a Carpenter.* New York: PowerKids Press, 1999.

Gibbons, Gail. *How a House Is Built.* New York: Holiday House, 1996.

Klinting, Lars. *Bruno the Carpenter.* New York: Holt, 1996.

Websites

B4UBuild.com Stuff 4 Kids
http://www.b4ubuild.com/kids

Build a Neighborhood
http://pbskids.org/rogers/R_house/build.htm

Glossary

architects: people who design buildings

blueprints: plans that show how buildings should be built

community: a group of people who live in the same city, town, or neighborhood. Communities share the same fire departments, schools, libraries, and other helpful places.

framers: carpenters who build the frames of houses

lumberyard: a store that sells wood and other building goods

plumbers: people who put water pipes in buildings

supplies: things that are needed to do a job

tools: things such as hammers or saws that help people do a job. Carpenters use tools when they are building.

Index

Photo Acknowledgments

The photographs in this book appear courtesy of: © Bill Miles/CORBIS, front cover; © Patrick Bennett/CORBIS, p. 3; © Royalty-Free/CORBIS, pp. 4, 7; © John Burke/SuperStock, p. 5; © Ariel Skelley/CORBIS, p. 6; © John Gress/Reuters/CORBIS, p. 8; © Philip Gould/CORBIS, p. 9; © Todd Strand/Independent Picture Service, pp. 10, 11, 13, 15, 20; © Index Stock Imagery/AbleStock, p. 12; © George Disario/CORBIS, p. 14; © Stephen Agricola/Photo Network/Grant Heilman, p. 16; © Don Mason/CORBIS, pp. 17, 22; © Stock Image/SuperStock, p. 18; © Nana Twumasi/Independent Picture Service, p. 19; © Jeff Greenberg/The Image Finders, p. 21; © Robert Llewellyn/CORBIS, p. 23; © Peter Beck/CORBIS, p. 24; © Jim Craigmyle/CORBIS, p. 25; © PhotoDisc Royalty-Free by Getty Images, p. 26; © Tom & Dee Ann McCarthy/Corbis, p. 27; © Brand X/SuperStock, p. 29.